797,885 Books
are available to read at

Forgotten Books

www.ForgottenBooks.com

Forgotten Books' App
Available for mobile, tablet & eReader

ISBN 978-1-334-18327-0
PIBN 10685001

This book is a reproduction of an important historical work. Forgotten Books uses state-of-the-art technology to digitally reconstruct the work, preserving the original format whilst repairing imperfections present in the aged copy. In rare cases, an imperfection in the original, such as a blemish or missing page, may be replicated in our edition. We do, however, repair the vast majority of imperfections successfully; any imperfections that remain are intentionally left to preserve the state of such historical works.

Forgotten Books is a registered trademark of FB &c Ltd.
Copyright © 2015 FB &c Ltd.
FB &c Ltd, Dalton House, 60 Windsor Avenue, London, SW19 2RR.
Company number 08720141. Registered in England and Wales.

For support please visit www.forgottenbooks.com

1 MONTH OF FREE READING

at
www.ForgottenBooks.com

By purchasing this book you are eligible for one month membership to ForgottenBooks.com, giving you unlimited access to our entire collection of over 700,000 titles via our web site and mobile apps.

To claim your free month visit:
www.forgottenbooks.com/free685001

* Offer is valid for 45 days from date of purchase. Terms and conditions apply.

English
Français
Deutsche
Italiano
Español
Português

www.forgottenbooks.com

Mythology Photography **Fiction**
Fishing Christianity **Art** Cooking
Essays Buddhism Freemasonry
Medicine **Biology** Music **Ancient Egypt** Evolution Carpentry Physics
Dance Geology **Mathematics** Fitness
Shakespeare **Folklore** Yoga Marketing
Confidence Immortality Biographies
Poetry **Psychology** Witchcraft
Electronics Chemistry History **Law**
Accounting **Philosophy** Anthropology
Alchemy Drama Quantum Mechanics
Atheism Sexual Health **Ancient History**
Entrepreneurship Languages Sport
Paleontology Needlework Islam
Metaphysics Investment Archaeology
Parenting Statistics Criminology
Motivational

UNITED STATES DEPARTMENT OF AGRICULTURE

BULLETIN No. 867

Contribution from the Bureau of Plant Industry
WM. A. TAYLOR, Chief

Washington, D. C.   PROFESSIONAL PAPER   September 3, 1920

THE CASTOR-OIL INDUSTRY.

By J. H. SHRADER, *formerly Chemical Technologist, Office of Drug, Poisonous, and Oil Plant Investigations.*

CONTENTS.

	Page.		Page.
The source of castor oil	1	Properties of the oil	31
Trade and commerce	3	Uses of castor oil	35
The inspection or valuation of castor beans	7	Conclusions	40
Manufacture of castor oil	9		

THE SOURCE OF CASTOR OIL.

Castor oil is obtained from the seeds of the castor-oil plant (*Ricinus communis*), a member of the plant family Euphorbiaceæ, which also includes many other species the seeds of which yield fatty oil. The castor-oil plant is now either cultivated or found growing wild in most tropical countries and in the milder parts of the Temperate Zones. In warm countries it is a perennial and often attains the proportions of a tree, frequently reaching a height of 40 feet, but in colder climates it rarely grows 20 feet high and annually dies down with the approach of winter. This plant is extremely variable in size, color of stems and leaves, degree of branching, size and color marking of the seeds, and many other characteristics.

The seeds of the castor-oil plant are known as "castor beans" in English-speaking countries, and are often mistakenly supposed to belong to the bean family, to which they are in nowise related. Castor beans as they occur in commerce consist of a white, soft, nonfibrous kernel inclosed in a hard, thin seed coat. This is so brittle that unless care is exercised in handling, it is easily cracked and chipped off, exposing the kernel to the air, with attendant deterioration. This seed coat is usually mottled in appearance and varies in color from a creamy white through pink, golden yellow, green, and

182601°—20——1

brown to black. The beans vary in weight from 0.10 gm. to 1.22 gm., and in size from 7.5 mm. long by 50 mm. wide by 4.5 mm. thick to 23 mm. by 15 mm. by 8 mm., respectively. One cubic foot of beans of average size weighs about 37 pounds.

An analysis of a commercial sample of castor beans shows that the whole bean consists of 35 per cent of seed coat and 65 per cent of kernel, the oil content of these two parts being 10 and 62.9 per cent, respectively.

Analyses of 37 samples of castor beans imported from India, China, the West Indies, and South America show a range of oil content, determined by ether extraction, from 38.39 to 55.55 per cent, with an average value of 48.75 per cent. Similar assay of 50 samples of beans from the first American crop from imported seed in 1918 shows a range of 42.13 to 58.57 per cent, averaging 48.16 per cent. With present milling practices there is obtained about 15.6 pounds of No. 1 oil per 46-pound bushel, 4.1 pounds of No. 3 oil, and 25.5 pounds of pomace, with a shrinkage per bushel of about 0.8 pound.

Castor beans contain a poisonous principle which renders them extremely dangerous when eaten. Also picking off the seed coats with the fingers has been followed by inflammation and soreness under the finger nails. Persons who have weak eyes or who are subject to hay fever may suffer serious inconvenience, to say the least, following exposure to the dust from ground dried castor pomace

The common belief that the coloring matter occurring in castor oil extracted by the solvent process is introduced through the so-called germ has not been borne out by experiments. On the contrary, the color seems to be mostly derived from the seed coats. It is observed that decorticated kernels yield a light-colored oil when extracted by volatile solvents, while the whole bean or the seed coats alone yield progressively more highly colored oils. But the coloring matter is evidently not a single simple substance, for upon extraction with different solvents different colors are obtained. Benzol gives a green-colored oil, gasoline a yellowish oil, while methyl ethyl ketone, acetone, ethyl alcohol, and methyl acetate all yield a red-colored oil which slowly turns to dark brown on exposure to the air.

Castor beans also contain the enzym lipase, which is extremely active in hydrolyzing the oil to glycerin and free fatty acid. This action does not occur so long as the seed coat remains intact and protects the kernel from exposure to the air, but when the seed coat is broken the enzym quickly begins its hydrolytic action. Since broken and damaged beans when crushed yield a highly colored acid oil, care must be exercised in hulling and shipping to avoid breaking the seed coats.

Another factor contributing to the high acidity of castor oil is what are known as black beans. These are rancid beans, the kernels of which have turned brown. Table I presents a comparison of the acidity of carefully selected beans with that of a mixture containing a known number and weight of black beans.

TABLE I.—*Comparison of the acidity of carefully selected castor beans with that of a mixture containing a known number and weight of black beans.*

Source.	Black beans in mixture by number.	Acidity (as oleic acid).	Source.	Black beans in mixture by number.	Acidity (as oleic acid).
	Per cent.	Per cent.		Per cent.	Per cent.
Miscellaneous imported lot	0	0.9	Brazil	0	1.4
	26	2		15	4.6
	50	7.7		24	3.3
	75	6	Bombay	0	1.2
	100	13		8.2	1.5
Do.	0	1.1	Haiti	0	.6
	16	2.4		15	1
	50	8.1	Venezuela	0	1
				10.7	2.1

It is interesting to note that in no case, even when all the beans were perfectly whole (controls), did the acidity as determined by titrating run below 0.6 per cent. Consequently, since at least a small amount of acid is invariably present, an acid determination of carefully selected sound beans would give a correct idea of the degree of freedom from acidity possible were all the beans equally sound. This percentage could not, of course, be realized in actual commercial practice, but it would show the standard toward which to work

TRADE AND COMMERCE.

The normal annual consumption of castor oil in the United States is more than 2,000,000 gallons, almost all of which is produced by our own crushing plants. The average annual importation of castor beans for the five fiscal years ended June 30, 1917, was about 834,000 bushels, while the importations of oil as such were comparatively insignificant.

The castor beans and castor oil of commerce come chiefly from India, China, the West Indies, and South America, with India producing by far the greatest quantity. In 1913 India exported 954,495 gallons of oil, while in 1917 this trade had increased to 1,723,463 gallons,[1] approximately 80 per cent of which was exported to the United Kingdom, 9 per cent to New Zealand, and 8 per cent to Australia. The center of the oil industry in India is Madras, which produces about 60 per cent of the total. Before the war the center of the industry was Bengal, which produced about 90 per cent of the castor oil exported. Table II gives the total exports of castor beans and castor oil from India for the years 1911–12 to 1917–18.

[1] These figures were supplied by the Far Eastern Division of the Bureau of Foreign and Domestic Commerce, February 21, 1919.

TABLE II.—*Exports of castor beans and castor oil from India for the years 1911–12 to 1917–18, inclusive.*[1]

Year.	Beans.	Oil.	Year.	Beans.	Oil.
	Tons.	Gallons.		Tons.	Gallons.
1911–12	120,194	1,404,803	1915–16	87,950	1,451,655
1912–13	110,630	954,495	1916–17	92,447	1,723,469
1913–14	134,888	1,007,001	1917–18	86,100	2,086,038
1914–15	82,815	898,269			

[1] Castor oil for aircraft engines. *In* Jour. Soc. Chem. Indus., v. 38, no. 2, p. 20R–21R. 1919.

Approximately 95 per cent of the castor beans exported were shipped out of Bombay. England took 4 per cent, the United States 23 per cent, France 18 per cent, Italy 47 per cent, and all other countries 8 per cent.

TABLE III.—*Imports of castor beans into the United States from various sources during the fiscal years 1910 to 1918, inclusive.*[1]

Country.	\multicolumn{9}{c}{Imports of castor beans.}
	1910	1911	1912	1913	1914	1915	1916	1917	1918
France....................bushels..	2	2	1		1		22		
Germany................do....	2		108		3				
Italy......................do....	12	14	2	5	27	21			
Netherlands.............do....	1	1	3						
United Kingdom........do....	503,773	368,966	135,137	43,392	105,390	163,267	223,208	91,007	607
Canada..................do....		2	2					1	
Cuba.....................do....	1,556								
Dutch West Indies......do....	5,729	8,165	984	1,669	3,136				1,587
Santo Domingo..........do....		102	3						
Brazil....................do....	158,351	122,918	44,633	8,961	10,049	9,744	7,887	94,148	169,312
British Guiana..........do....	28								
Venezuela................do....	2,839	1,930				789		540	14,367
British India............do....	53,707	242,935	777,074	833,720	905,946	750,039	839,282	517,711	620,480
Jamaica (British West Indies), bushels.................				41		131	59		
Turkey in Europe.....bushels..						2,229			
Other British West Indies, bushels.................						13			32
China..................bushels..						3		1,542	28
Dutch East Indies......do....						3,610	685	14	263
Japan....................do....						5			10,284	124,793
Honduras................do....									2
Mexico...................do....								39	2,541
Dominican Republic....do....								58	608
Haiti....................do....								1,242	80,687
Argentina...............do....								13,275	9,067
Colombia................do....								400	771
Japan (Chinese leased territory)..............bushels..								37,887	13,426
Hongkong...............do....							2		4,295
Costa Rica..............do....							6		212
Panama.................do....									256
Trinidad and Tobago...do....									21
Peru....................do....									824
Uruguay................do....									1
British South Africa....do....									79
Total........{do.... {tons..	726,002 16,698	745,035 17,136	957,986 22,034	887,747 20,418	1,030,543 23,702	924,604 21,266	1,071,963 24,655	766,857 17,638	1,144,014 24,012

[1] These figures are taken from "General imports" in Commerce and Navigation of the United States and represent the bushel as containing 50 pounds of beans.

Castor beans grow quite abundantly all over China. Reports from United States consuls show that the total exportation of castor oil from seven ports in 1913 was 3,677,266 pounds, while in 1916 the total for these same ports reached 9,108,133 pounds. These ports in the order of their importance, with percentages, are as follows: Tientsin, 45; Shanghai, 19; Newchwang, 17; Kiaochow, 11; Dairen, 5; Hankow, 3; Canton, negligible.

Owing to the campaigns inaugurated by the Allies to increase the world's production of castor beans, notable quantities of beans and some oil have been entering the world's commerce from Central America, Africa, Japan, and the Malay Archipelago. Table III gives the imports of castor beans from various sources into the United States during the fiscal years 1910 to 1918, inclusive.

The castor beans and castor oil imported for consumption during the fiscal years 1910 to 1919, inclusive, minus the quantities exported, are presented in Table IV. Inasmuch as there was during this period practically no American commercial production of castor beans, these figures fairly represent our trade in this commodity. It was only during 1918 that there was really a commercial American crop, and this was due to the demand for oil for aircraft lubrication.

TABLE IV.—*Castor beans and castor oil imported for consumption, minus the quantities exported, during the fiscal years 1910 to 1919, inclusive.*[1]

Fiscal year.	Castor beans (bushels).	Castor oil (gallons).	Total as oil. Gallons.	Total as oil. Pounds.	Fiscal year.	Castor beans (bushels).	Castor oil (gallons).	Total as oil. Gallons.	Total as oil. Pounds.
1910...	752,374	....	2,106,647	16,853,176	1915..	924,605	63,005	2,651,899	21,215,192
1911...	790,241	....	2,212,675	17,701,400	1916..	1,071,969	253,077	3,354,590	26,036,720
1912...	978,257	6,971	2,746,090	21,968,720	1917..	767,019	323,703	2,471,456	19,770,848
1913...	824,573	5,241	2,314,045	18,512,360	1918..	1,262,130	1,175,064	4,709,028	37,672,224
1914...	1,043,928	189,583	3,112,591	24,900,728	1919..	628,312	....	1,759,274	14,074,192

[1] These figures are taken from "Imports for consumption" in Commerce and Navigation of the United States, and represent the bushel as containing 50 pounds of beans, yielding 45 per cent of oil, or 22.5 pounds, which is equivalent to about 2.8 gallons of oil.

From November, 1918, to June, 1919, inclusive, the Government plant at Gainesville, Fla., was crushing American-grown castor beans. It crushed 211,000 bushels (of 46 pounds), yielding 463,000 gallons of oil, or 3,730,000 pounds. This quantity, of course, is to be added to that produced by the regular castor-bean crushing industry from imported beans, reported above, to show the total consumption and stocks for the fiscal year 1918–19. In the winter of 1917–18 about 6,000 bushels of imported castor beans were used for planting.

The rate of duty on castor beans from August 5, 1909, to October 4, 1913, was 25 cents per bushel; subsequent to that date it was 15 cents. The duty on the oil for corresponding dates was 35 and 12 cents per gallon.

In 1912, 88 per cent of the beans imported entered at New York, 8 per cent at Grand Rapids, and 4 per cent at Boston. Table V shows the relative quantities of beans imported from various sources for the years 1912, 1914, and 1918.

TABLE V.—*Relative quantities of castor beans imported from various sources during the years 1912, 1914, and 1918.*[1]

Source.	Relative importations of castor beans (per cent).		
	1912	1914	1918
United Kingdom	14.0	10.0	0.06
India	81.0	87.9	60.00
South America	4.6	1.0	19.00
West Indies	.1	.3	8.00

[1] Calculated from Table III.

In 1918 the American castor-bean crop was estimated at about 5,750 tons (250,000 bushels), or about one-fourth of the normal domestic consumption. The sudden cessation of hostilities, however, together with the great impetus given to castor-bean cultivation throughout the world, has resulted in an overburdened market, with consequent tendencies to lower the price and unload foreign stocks in the United States. This would certainly have a tendency to cripple the crushing industry here were it not for the high quality of oil produced. In fact, the best grade produced in this country is not excelled by any foreign producer, while some of our No. 3 grade is as good as much of the imported first-grade oil. However, there is a growing tendency among foreign vegetable-oil producers to erect crushing mills near the areas of production of oleaginous materials, which would certainly seem in course of time to affect our supplies of raw material, with the consequent crippling of our mills.

The relieving feature of this development is the intimate knowledge gained of the possibilities for creating a permanent American castor-bean industry. In the past this has been an actuality. Until about 1900 relatively large quantities of castor beans were raised in this country,[1] chiefly in Oklahoma, Kansas, Missouri, and Illinois, but in the face of successful foreign competition the possibilities for development of the American crop disappeared. The great campaign of castor-bean growing inaugurated in 1917 by the Bureau of Aircraft Production has resulted in gathering considerable information concerning the growing of castor beans, such as yields per acre in different parts of the country and cost of handling, and we are now in a good position from the standpoint of knowledge of farming conditions to adopt intelligently whatever measures may be necessary to meet foreign competition. Most of the details of seed

[1] Yearbook, U. S. Dept. of Agriculture, 1904, p. 295.

selection and methods of planting, cultivating, and harvesting are being worked out. But the farmer who would raise castor beans as a crop will have to be shown that he can receive more money per acre than he is receiving from his present crops before there will be a satisfactory home production of castor beans. Cost, yield, market, and profit are the determining factors.

The relative consumption of castor beans in the great vegetable-oil producing countries for the years 1911 to 1913 is shown in Table VI. Exports are subtracted from imports, all figures being reported in the respective commerce returns.

TABLE VI.—*Consumption of castor beans in the great vegetable-oil producing countries for the years 1911, 1912, and 1913.*

Countries.	Consumption of castor beans (bushels).		
	1911	1912	1913
United States	790,000	978,000	824,000
United Kingdom [1]	1,928,000	1,856,000	2,132,000
France [2]	460,360	663,520	1,006,280
Germany [3]	336,760	377,720	419,160

[1] Annual return of trade. [2] Tableau général du Commerce et de la Navigation.
[3] Auswärtige Handel-Statistik des Deutschen Reichs.

The United Kingdom exported much of its home-crushed oil, leaving 3,624,000 gallons (14,495 tons) of oil consumed in 1913, figured by subtracting the exports from the imports, whether oil as such or calculated on a 45 per cent yield from the beans. The American consumption in the same year, 1913, figured similarly, was 2,314,045 gallons (9,250 tons). (See Table IV.) The figures for France and Germany only very roughly follow consumption, as the customs returns include this commodity with others. Hence the figures represent a maximum never attained by castor beans alone.

THE INSPECTION OR VALUATION OF CASTOR BEANS.[1]

The castor beans of commerce are bought on a standard form of contract of the Linseed Oil Association in New York City.[2] If 3 per cent of impurities, or less, is present in any lot of beans no deduction or reward for impurities is made, but if more than 3 per cent is present a deduction is made for all over such a figure. These impurities include hulls, sand, and pebbles, sometimes stones 2 or 3 inches in diameter, occasionally extraneous seeds, and even foreign money. Beans from India usually run well under 3 per cent of impurities,

---

[1] Contributed by A. C. Goetz (Capt. A. S. Sig. R. C.), formerly in charge of the Chemical Supplies and Materials Section of the Production Division, Inspection Department, Bureau of Aircraft Production.
[2] The Linseed Oil Association of New York City is an association of crushers, brokers, commission men, etc., dealing in linseed oil, castor beans, and other oleaginous materials.

but those from South America and the West Indies as a rule contain more dirt and trash than the beans from India. Shipments may also contain immature beans, decorticated beans, black beans, and broken beans. The decortication is done in the thrashing. Black beans (usually yellow, brown, or black) are those whose kernels have become discolored as the result of the beans having been wet. In imported lots, broken and decorticated beans as a rule are not excessive, but if the percentage of such black beans does run high, a claim is made, about 5 per cent being the maximum allowable figure. Broken and decorticated beans are not classed as dirt or impurities, but if the percentage seems excessive the claim for reduction is made independent of the percentage of trash. Black, broken, and decorticated beans have the effect of increasing the acidity of the oil produced therefrom.

Shipments of castor beans are sampled by commercial samplers. This is done by means of a tryer, samples being taken from 5 per cent of the bags in a lot, but no sample should be less than 50 pounds. The sample is divided into as many smaller ones as are necessary, one for analysis, one for record or file, and one for the purchaser or seller if he so desires. After the sample is obtained it is entirely in the hands of the Linseed Oil Association, which association makes the necessary records and arranges for an analysis by its official analyst. This analysis is accepted by both buyer and seller. Any dispute arising regarding the percentage of impurities or the quality of the beans is usually settled by the interested parties themselves, although the custom of the trade of having a committee or referee appointed is sometimes resorted to. As a rule, no beans are refused outright; usually any lot of beans will be taken at some price. If resort is taken to a committee or referee, the disputes as to quality are not settled in this country but are settled by the Oil Seeds Association of London, England, which considers the fair average quality of the season's crop. Such instances, however, are not frequent.

The weight per standard bushel is another factor indicative of the quality of castor beans. This weight varies with beans from different localities. Considering beans from the same locality, the heavier the weight per bushel the better the quality of the beans.

Inasmuch as castor beans are bought for oil making, it would seem that the logical way to judge them would be on the quantity and quality of the oil actually contained in the beans, in addition to the test for impurities, all of which the buyer should know before purchasing. He would then not be obliged to judge the oil in the beans on only an approximation. Such a method of purchase and inspection might require a longer time to complete tests, but they should be completed within 24 hours from the receipt of the sample, which does not seem impracticable. Oil-bearing seeds and material are not yet bought on the above-suggested basis, but sooner or later such will doubtless be the practice. The only objection to this seems to be the desire of those interested not to change present methods, fearing that complications would be added to transactions unfavorable to growth in trade.

MANUFACTURE OF CASTOR OIL.

CLEANING.

The castor beans of commerce come mixed with such amounts of trash that it is absolutely necessary to remove this at the earliest stages of the operation. An ordinary grain cleaner is used, with special perforations in the screens for castor beans. These cleaners are a combination of screens and fans, which remove the straw, hulls, and dirt in successive operations of sieving and fanning and deliver the cleaned beans. Such machines vary greatly in size, with attendant capacities of 30 to 1,200 bushels of beans per hour.

DECORTICATION.

Owing to the peculiar structure of the castor bean, its milling technology has developed along lines peculiar to itself. Several special machines have been built to decorticate castor beans. The process depends on cracking the brittle seed coats between rolls set so as to exert a cracking pressure rather than a crushing and mangling one. The beans with their broken seed coats are dropped on a shaking screen which serves to shake the kernels out of the adhering seed coats and otherwise loosen up the mass. As the charge falls over the end of the shaker, a current of air blows out the seed coats, while the kernels fall into a hopper below.

It is very difficult to make a perfect separation of hulls from kernels by the mechanical decortication of castor beans as they are usually delivered from the warehouse. One of the factors contributing to this difficulty is the irregularity in the size of the beans, the small ones dropping between the rolls unaffected, while the large ones are crushed. This requires a grading of the beans, which of course can readily be done mechanically.

Such decortication would be all the better effected if the beans were given a preliminary drying, such as could easily be accomplished in the heaters described later. With regard to the question of added moisture for optimum pressing conditions, see under "Heating."

The advantages of decorticating consist in giving a somewhat lighter colored oil and at the same time removing the constituents which wear out the equipment. The literature persistently reports that the Italians produce a very superior medicinal grade of oil, which is practically tasteless, by pressing beans which have been decorticated or dehulled and carefully picked over by hand to remove the bits of adhering hull. On the other hand, the lack of sufficient fiber necessitates in most operations the addition of a so-called binder to prevent the troublesome squirting of the meats from the presses and at the same time prevent the introduction of excessive meal into the oil. The disadvantages so greatly outweigh the advantages that this practice is not followed to any great extent either here or abroad. (See under "Solvent extraction.") A decorticating machine is illustrated in figure 1.

HEATING.

After cleaning, the beans are conducted to the heaters. In contradistinction to other vegetable oleaginous materials, castor beans can not be ground and tempered as can flaxseed, peanuts, soy beans, cottonseed, and copra. The nonfibrous character of the kernel with its attendant high oil content so gums up the equipment that further grinding is difficult. In addition to these technical difficulties, the bean contains a very active lipase or fat-splitting (saponifying) enzym, which very quickly sets free an excessive amount of fatty acid. Accordingly, every effort is made to deliver the beans to the presses as whole as possible. The reason they are heated at all is to render more mobile the naturally heavy viscous oil. Probably even this would not be done if such treatment served materially to impair the quality of the oil. The temperature to which the beans are raised varies from 100° to 120° F.

FIG. 1.—A castor-bean decorticator.

To test the effect of heating undamaged beans, a sample from Santo Domingo was heated for 2½ hours at 70° C. (158° F.). It was then carefully picked over to eliminate any black beans, and cold pressed without any further heating in a laboratory cage press at 1,500 pounds applied pressure per square inch. A sample of unheated beans wa pressed as a control. In each case the first runnings were discarded

CASTOR-OIL INDUSTRY. 11

after "wetting" the equipment. The acidity of the oil of the control (unheated) beans was 0.37 per cent, while that of the heated beans was 0.31 per cent, showing no deleterious action due to such heating. The color of the oil in both cases was the same.

A type of heater used in this country for castor beans and known as a grain drier is illustrated in figure 2. This equipment, including the accompanying racks (figs. 3 and 4), is constructed entirely of galvanized steel plates, pressed into the desired shapes and cleated or riveted together in sections, which are assembled and bolted together to build up any desired capacity. The bean-holding compartments consist of a series of vertical racks made up of horizontal steel shelves (pitched bottoms) attached to vertical steel plates. These shelves are staggered opposite each other in such a manner that beans entering the upper end of the racks will descend through a zigzag course between the shelves and from one shelf to another until they are stopped at the bottom by a series of slides operated by rockshafts and levers. They will then pile up vertically without overflowing or leaking from the sides of the racks until the entire height of the racks is again full of beans, forming vertical zigzag layers with both sides entirely open. The beans do not pack in these vertical columns, for the reason that each shelf bears the load of the beans resting directly upon it and the weight is distributed equally throughout

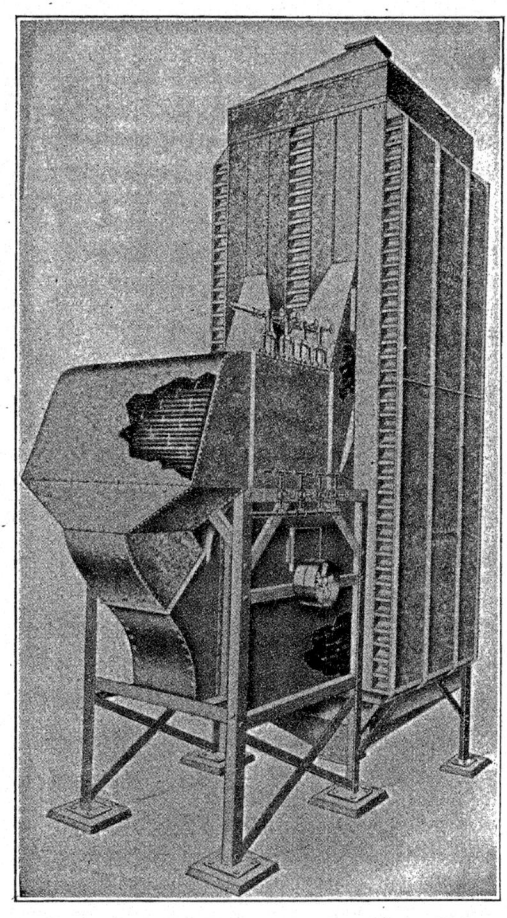

FIG. 2.—A small assembled heater. It is to be noted that the lower cooling section is discarded when used for heating castor beans.

the height of the racks. By adjusting the slides at the bottom, controlled by levers, the discharge may be made continuous or intermittent. Usually the latter practice is followed.

The heating is effected by rising columns of hot air. This air is drawn in by a fan at the bottom and side of the bean-holding equipment and passed over hot closed steam pipes. To prevent the beans from drying out, and also to increase the heating effect of the hot air,

FIG. 3.—Bean racks, showing steel uprights and shelves, saddles, and slide levers. Every surface on which beans are supported is inclined so the beans and dust will flow out when the slides are open, leaving the racks clean. These racks are of smooth galvanized steel and are fitted up closely, without bolts.

FIG. 4.—Section of bean rack showing it (at the left) discharging at intervals and (at the right) in continuous flow. By throwing the small levers at the bottom of the racks the method of discharge is instantly changed from one to the other. The upper large lever operates the slides for discharge at intervals, but is not used when the beans are flowing continuously.

a jet of live steam is sometimes introduced into the hot-air stream as it emerges from the nests of steam pipes. It is found that an air temperature of about 180° F. will raise that of the beans when delivered to the machines to about 110° F.

'Although the above is a method of procedure now followed in commercial practice, it might be well to consider the following facts with regard to improved operation. It is generally conceded in expeller practice that better oil yields are obtained when appreciable quantities of moisture are present. These quantities vary with the character of the material as well as with the way in which the moisture is held, that is, whether or not the moisture is internal and evenly distributed throughout the kernels or oil-bearing particles, on the one hand, or whether, on the other hand, the moisture is distributed over the surfaces of the particles.

To obtain the former condition, the beans are heated in an atmosphere highly charged with moisture in order to prevent drying out, while in the latter case they should be heated in a dry atmosphere with ample facilities for the removal of moisture. In order, then, to add the proper amount of moisture for optimum crushing conditions, a jet of live steam should play into the descending current of material just before it enters the expeller. Such operation has the added advantage of (1) increasing the quantity of oil obtained and (2) securing the economical advantage of using the latent heat of steam to effect the heating of the beans just previous to their delivery to the expellers, thus economizing in the heating operation. Such procedure, of course, gives a dry interior to the oleaginous material and a wet surface.

With further application to improved castor-bean technology, it is evident that this drying effect on the beans would improve and greatly simplify decortication by rendering the seed coats more easily separated from the kernels.

The heaters are located close to the presses in order to prevent loss of heat and to avoid heating the beans for a longer time and to a higher temperature than is absolutely necessary. The lower the temperature to which the beans are subjected and the less the moisture in the oil, the better is the finished product. Heat darkens the oil and moisture increases the acidity.

PRESSING.

In the past, plate presses and cloths were used for castor-bean crushing. On account, however, of the nonfibrous character of the beans with their attendant property of "creeping" under pressure, the cloths would break and the meats would squeeze out. When castor beans are crushed in such hydraulic presses to give a cold-drawn oil, it is stated that a yield of about 17 pounds of oil to the bushel is obtained. Regrinding the cake and heating to 180° to 200° F. for a second pressing yields an additional 3 pounds of oil, leaving about 5.5 per cent of oil in the cake. These figures probably refer to the 46-pound bushel, but the quality of the oil is not up to that from cage presses or expellers. Pressures great enough to yield 17 pounds of No. 1 oil per 46-pound bushel are liable to introduce objectionable impurities. However, with pressure sufficient to yield 16 pounds of No. 1 oil, this oil need not be discounted.

This method has been abandoned for that of the cage press. Essentially this latter is an ordinary hydraulic plate press with the plates removed to accommodate an iron box (or cage) with rectangular or cylindrical cross section, open at both ends, and with the walls perforated over their entire surface. This cage is set vertically in the press in place of the removed plates, while the ram head just fits the cage and rises into it. A headblock above the cage and attached to the press frame likewise fits into the cage from above and supplies the resistance against which the ram operates. Other cages consist of vertical bars set closely and clamped by heavy rings to resist the bursting pressures.

The method of filling the cages varies with different types of presses. Some cages are filled on a so-called charging press and are transferred with their contents to the finishing presses, while others press in only one operation. In any case, when charging, the cages remain in position in the press frame with the ram raised to within a short distance of the top of the cage. The cage is filled with beans level with the top and a plate is laid on, closely fitting the walls. Sometimes a mat is laid on the plate. The ram with its charge then descends a given distance, leaving a space equal to the first one. This is filled level with beans and another plate added. The process is continued until the ram has been lowered to the bottom of the cage, leaving the latter filled with layers of beans separated at given distances by plates. The number of such plates may vary from 6 to 50. Other types have two horizontal rails in front of the cage whereby the latter can be pulled out from the press frame and either filled from above by dropping in measured quantities of beans separated by plates or discharged of its press cake by dumping below into a hopper or conveyor. The cage is then shoved back within the press frame for pressing.

Whatever preliminary method is used for charging the cage, the subsequent pressing operations are quite similar. The headblock is placed in position so as to engage in the top of the cage. As the ram rises, it compacts the charge of beans and presses it against the headblock, thus developing increasing pressures which cause the oil to start. Maximum yields and a better quality of oil are obtained by so adjusting the rate of rise of the ram that the pressures are developed slowly rather than suddenly or unevenly. Pressures on the ram of 4,000 to 6,000 pounds per square inch are now in operation, and some foreign mills are being built to apply 8,000 pounds to the square inch. The duration of the operation varies from 15 minutes to 1 hour, depending on whether a large tonnage of beans is being handled with a higher content of oil in the cake or a less tonnage with a correspondingly greater yield of oil per bushel of beans. Typical cage presses are illustrated in figures 5 to 10.

CASTOR-OIL INDUSTRY. 15

In figure 5 the press at the right shows the overhead plunger rolled back and the cage ready to receive the crushed material; the press at the left shows the plunger in position to receive the pressure. During the pressing operation metal shields are placed around the

FIG. 5.—A cage press inclosed in a shield (at left) to prevent the squirting of the oil. Overhead plunger rolled back to permit filling the cage (at right).

retaining cages for the purpose of preventing the oil from squirting and to conduct it to the pan below.

Figure 6 shows an ordinary plate press which has been converted into a cage press by removing the plates and substituting a small specially constructed cage to fit within the press frame, holding about 1 bushel to the charge. A frame has been added in front of the cage, on which to slide out the latter for filling and dumping.

Figure 7 shows a filling press, the object of its use being to fill a cage, tamp the contents, and thus conserve pressing space in the operations which follow. Otherwise, the initial shrinkage of the batch due to the application of high pressure would result in much loss of valuable pressing space. Figure 8 shows a discharging apparatus used in removing the cakes from the press, a comparatively slow operation, the object of its use being to avoid diverting heavy pressures from their intended use. Figure 9 shows a finishing press for the heavy pressure for the production of oil. Figure 10 shows a filling, finishing, and discharging press, where all operations are

FIG. 6.—A cottonseed plate press converted into a cage press. Cage under ram, ready to receive pressure, at left; press, showing cage out ready to be charged, at right.

performed in the same press. Figure 11 shows a carriage with the cage used in transporting a cage between the various presses.

After pressing, the cakes are removed from the retaining cages by rolling back the overhead plungers and applying the pressure. The cages are bolted in a stationary position to the press pan and the press ram works up through the cage, forcing the pressed cakes out at the top, from which point they are removed by the operator.

Castor-bean cake is very different from that of other oilseeds. In the latter, the cakes are as firm and low in oil as any produced in plate presses. In fact, the newer developments of cage presses are producing cakes lower in oil than those obtained by any other pressing operation. In the case of castor beans, however, the cakes are not cohesive and dry, but readily crumble and fall to pieces.

CASTOR-OIL INDUSTRY. 17

Usually they contain from 12 to 20 per cent of oil, according to the length of the pressing cycle and the pressure applied. These high percentages of oil in the cakes are caused by the presence of the seed coats, which form little cups throughout the mass and prevent ready draining of the oil and crushing of the kernels. So incomplete is the crushing that the kernels often retain their original shape. On the other hand, it has been stated that the application of too great a pressure produces oil which precipitates an albuminlike product on standing. Thus, like most commercial operations, there

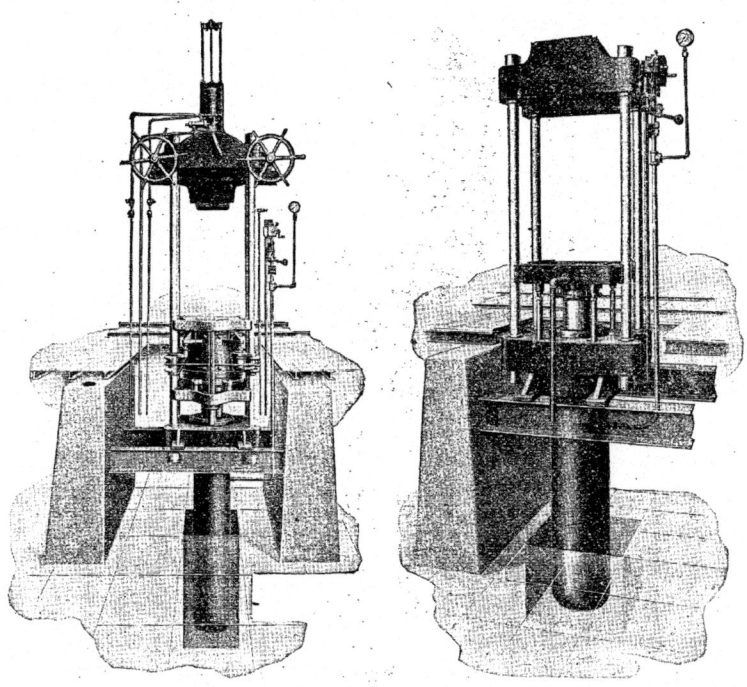

FIG. 7.—A cage filling press.   FIG. 8.—A cage discharging press, used to remove the cakes.

comes a point where a satisfactory balance is struck between quantity and quality, beyond which the operator sacrifices profits. Good practice yields about 15.6 pounds of No. 1 cold-pressed oil to the bushel of beans. with 4.3 pounds remaining in the cake as No. 3 oil, assuming 46 pounds of beans to the bushel and an oil content of 45 per cent. Such large percentages of oil in the cake, together with the high prices of the oil, warrant treatment of the cake for the recovery of this oil. This, of course, can be accomplished only by solvent extraction.

182601°—20——3

These cage or curb presses are made in various sizes, with capacities varying from 100 pounds of beans per hour to about 1 ton. The curbs correspondingly vary in diameter from 10 to 19 inches and in height from 2½ to 9 feet. An actual installation of such equipment consists of a heating kettle provided with a steam jacket and stirrer immediately over the press, so that the heated contents of the kettle can be discharged through an opening in the bottom directly into the cage. These beans are heated to about 90° F. and pressed to the desired degree. The resulting cakes contain only about 10 per

FIG. 9.—A cage finishing press for the heavy pressure necessary for the production of oil.

FIG. 10.—A cage filling, finishing, and discharging press, where all operations are performed in the same press.

cent of oil. Four such presses operated by two men have capacities of 500 pounds at each pressing, or an hourly total of 2 tons for the four presses with two pressings; 3 tons with three pressings per hour.

EXPELLING.

The other method used in modern oilseed milling which has been applied to castor-bean pressing is that of expelling the oil in Anderson oil expellers. These machines are built around a horizontal cage or barrel, about 6 inches inside diameter and 33 inches long, which consists of bars about 0.025 inch apart at the feed end, 0.015 inch at the middle, and 0.025 inch at the discharge end, all bound together by massive clamping bars. A horizontal screw rotating on the axis of the cage carries the charge of oilseed forward and discharges it at the other end of the cage over a cone, which may be moved in or out, according as greater or less pressures are sought. The charge of oil-bearing material is introduced through a hopper into the cage or barrel and the slowly rotating worm engages it and carries it forward. As the charge reaches the small aperture around the cone it encounters more resistance, which operates to build up pressures within the barrel. These pressures are regulated by varying the size of the aperture between the cone and the walls of the barrel. The farther the cone is moved into the barrel, the smaller is the aperture through which the cake may be discharged. Inasmuch as the screw moves forward at a constant rate and thus delivers a continuous quantity of material to the barrel, it follows that the oil yield is readily controlled by simply adjusting the cone and varying the size of the discharging aperture. For ordinary materials these expellers can be adjusted to yield cakes containing as low as 5.5 per cent of oil.

These machines are very generally used for peanut and copra crushing and for many other oleaginous materials, all of which, however, must have such fiber content as to present sufficiently effective binding properties to prevent squirting of the meats through the interstices of the bars. There is, of course, great wear of the parts, but since these are standardized they can readily be supplied. It has been stated that with properly cleaned material and with right care of the machine the upkeep is no greater than the cost of the cloth in plate presses. An average for more than seven prewar years shows the upkeep to have been about $100 for each machine per annum.

FIG. 11.—A cage carriage.

Figure 12 represents a battery of such machines set up somewhat as they appear in good factory installation. Manifestly, the small labor charge attendant on their operation is much in their favor, together with the cleanliness and mechanical simplicity of their action. One man can operate the whole battery, and even more.

FIG. 2.—Battery of expellers, with pump, filter press, and oil tank.

Expellers for castor beans should be designed similar to those used for copra; other types are not as satisfactory. Such machines have three worm flights on the pressing screw, whereas those intended for crushing other more fibrous and lower oil-bearing materials have only two flights.

Castor beans being low in fiber can not readily be handled in the expellers when excessive pressures are applied, owing to the large quantity of meal which accompanies the oil. If subsidiary equipment is used for removing the meal, this objection is minimized. However, inasmuch as such high pressures are accompanied with the generation of appreciable quantities of heat, which serve to darken the oil, and also since the cake is to be extracted for the residual oil, the market for which is only very slightly below that of pressed oil, good practice applies comparatively little pressure, which results in the production of a cake about seven-sixteenths of an inch thick containing about the same amount of oil as that produced by cage presses, namely, 12 to 15 per cent.

Because of the necessity for some binder to hold the cake within the machine and to preclude the squirting of the meal, it is impracticable to decorticate castor beans. However, if decortication is desired for any purpose, the necessary binder can be mixed with the kernels to hold them and to form a cake. To test this method, decorticated castor beans were mixed with 9 per cent of peanut hulls and expelled as usual. The resultant cake came out in very good form, and the oil was quite free from meal. Since decortication reduced the tonnage pressed by 35 per cent and the added fiber was only 9 per cent, it follows that a net reduction in tonnage of about 26 per cent was effected, resulting in approximately one-third greater expelling capacity.

The unsatisfactory results accompanying earlier attempts to apply the expeller to castor-bean oil manufacture are not hard to understand. The wearing parts of the expellers made 15 years ago were of cast steel and consequently were quite susceptible to the abrasive action of the hard siliceous seed coats. Since then, however, the wearing parts are of case-hardened steel, which presents an altogether different aspect to the abrasive action. The Government castor-oil mill at Gainesville, Fla., has an installation of 15 expellers which have been operating for several months and yet show actually less wear than is observed when used for expelling peanuts. Out of a possible linear movement of 2½ inches in which the cones can be moved to take up the wear, they have been taken up only half an inch.

A test batch of 1,701 pounds (37 bushels) of castor beans heated to 140° F. was expelled in a regularly installed factory expeller. The operation required 2.27 hours and yielded 727 pounds of oil and 920 pounds of cake. Based on the 46-pound bushel, this is equivalent to

2.46 gallons (19.7 pounds) of No. 1 oil, containing about 10 per cent of meal. The beans assayed 48.4 per cent oil. Inasmuch as there is normally about 2 per cent of trash with an additional loss of about 1 pound per bushel for shrinkage, it follows that the oil content of the cake was about 11.6 per cent. The No. 1 oil was agitated with 5 per cent fuller's earth and 2 per cent filtchar at 85° to 90° C. for 10 minutes, then for 15 minutes at 70° to 85° C., and filtered. A very light colored oil was obtained. The above pressure was too great, however, for satisfactory mill operation.

Messrs. Pinnock and Goetz, of the Bureau of Aircraft Production, expelled the oil from a batch of beans in the hull in order to test the possibility of obviating the expense of maintaining hulling stations scattered over the country, and also of assembling in one place all merchantable castor-bean products. By such practice all hulls are included in the residual pomace as fertilizer, the costs of hulling and attendant contractor commissions are eliminated, shipping costs are reduced by eliminating bagging, while an oil of lower acidity is obtained, owing to the protection of the brittle seed coats from cracking during handling. The great disadvantage, however, is that a green-colored oil is produced. For lubricating purposes this is of little moment, but in ordinary trade it is unacceptable. However, the oil can readily be bleached by refining it first with alkali and then bleaching with fuller's earth and carbon. From 258¼ pounds of beans in the hull, 79¾ pounds of oil and 166¾ pounds of cake were obtained. Since the original oil content of the beans was 34.5 per cent, it follows that the operation yielded about 92.75 per cent of the total oil. No oil was lost by absorption in the extra cake, owing to the fact that the hulls themselves contained 6.4 per cent of oil.

Unless proper care and skill are applied in using the expellers the machines produce an oil high in meal. To prevent this, it is necessary that the proper amount of moisture be present, and that the beans be heated. On starting a cold expeller, a thick mass of meal and oil is produced, but after running about half an hour the machine warms up and produces an oil with apparently no more meal than that from a hydraulic cage press. In a day's run of seven hours there is produced only about 1 quart of such meal by each machine. Oil made in expellers is certainly heated to higher temperatures than obtain in cage pressing. As it drips through the bars it has an average temperature of about 176° F., while that of the cake as it emerges at the discharge is about 133° F. So little deleterious effect is produced on the oil by this heat that it must be attributed to the extremely short duration of the exposure.

The expellers in the Gainesville plant have been regularly operating on 800 pounds, or about 17 to 18 bushels of beans an hour, but under forcing can do materially more. A test on an experimental expeller indicated a possible thousand pounds, or an hourly rate of 21 bushels. In the expeller plant at Gainesville, the machines were not set to yield more than about 15 pounds of oil per 46-pound bushel.

The relative quality of expeller castor oil compared with that of hydraulic pressed oil will be considered later.

SOLVENT EXTRACTION.

On account of the large proportion of residual castor oil left in the cake from the pressing or expelling operation, namely, from 12 to 20 per cent, or about 5 pounds per 46-pound bushel, solvent extraction of the cake is universally practiced in this country, following the methods generally applicable to oleaginous products. The units comprising a complete extraction plant consist of the extractor and the solvent-recovery still, the oil-finishing still, the condenser, the solvent and water separator, pumps, and storage tanks.

Two general types of equipment are used, namely, the stationary extractor and the rotary extractor. The former is primarily an English development, while the latter is American, having found extensive application in the extraction of garbage and other waste materials where the great problem is to handle in an economical manner bulky products with small oil content.

The iron stationary extractors vary in dimensions, but may have a diameter of 6 to 8 feet and a height of 10 to 12 feet, entirely inclosed, with, however, apertures for loading and discharging and also stirrers and vapor pipes. The extractor is provided with a false bottom, perforated over its entire surface with small openings, varying in size with the character of the material to be extracted. These openings may be over an inch in diameter, but if finer materials are to be extracted auxiliary plates with smaller perforations may be laid on.

In order to obviate channeling in castor-pomace extraction, attempts have been made to operate a stirrer, but owing to the tendency of such material to pack during extraction it has been found impracticable to operate such stirrers when the extractors are charged to anything like their capacity. Accordingly, recourse is had to auxiliary methods of breaking up channeling in extraction, as well as in "steaming off" the solvent in the final treatment. Another difficulty of stationary extractors is that there is a tendency for materials containing large quantities of nonfibrous albuminous material to pack on the bottom, which very effectively precludes the circulation of the solvent, so essential to the efficiency of extraction. Among the methods which have been used to obviate this packing tendency of castor beans is that of laying burlap between quarter-inch chicken wire over the floor, spreading thereon a layer of hulls, and then piling the pomace on this. Such an arrangement in conjunction with the introduction of the solvent from below insures the movement upward rather than downward of the albuminous products and other sedimentation and greatly minimizes the tendency to pack. Such an extractor may hold a charge of 10 tons with a period of operation of possibly 20 hours from the commencement of loading of one batch to that of the following batch.

The advantage of the stationary type of extractor is simplicity of installation and absence of more or less complicated machinery.

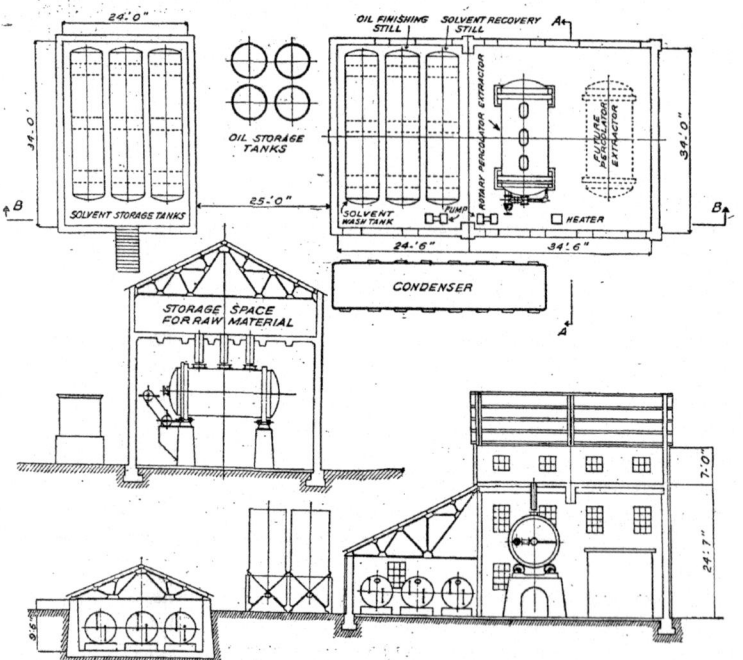

Fig. 13.—Sections of extraction plant.

The rotary extractor is finding increasing favor because of its ease of operation, efficiency of performance, and low labor charge. The central unit of the rotary extractor plant (fig. 13) consists essentially of a closed cylindrical tank, mounted horizontally upon riding rings and provided with a series of manholes along one side for the introduction and discharge of material. This tank is mounted upon trunnion rollers and is rotated by means of a girt gear which is attached to one end of the shell and driven by a set of geared countershafts. A false bottom covered with burlap extends along the entire length of the tank, thus leaving a clear flow for the solution of oil in the solvent. Such extractors are made in various sizes and may be from 5 to 8 feet in diameter and from 12 to 18 feet long. The solvent and steam are introduced through openings in the top; the saturated solvent and steam are drawn off at the bottom. Another type introduces steam and solvent through a journal bearing in the axis at one side and discharges steam and saturated solvent through a similar opening in the axis on the other side, first passing through a perforated false head which divides the discharge end of the tank, into a small compartment into which the filtered solvent collects. The smallest of such units will hold about 3 tons of oleaginous material to the charge, while the large ones may hold as much as 6 tons. The total period of treatment from the time of the commencement of loading back to the corresponding stage of the succeeding batch may be about 12 hours, enabling the units to be used twice in 24 hours. The channeling of the batch is obviated by rotating the tank at intervals, thus breaking up pockets and other deterrents of good extraction.

With the exception of the form of the extractor, the rest of the solvent plant is quite standard and simple. The solvent-recovery still may be provided with a steam jacket on the bottom with auxiliary heating coils, or with a calandria somewhat similar to that used with sugar pans (simulating an upright fire-tube boiler) where steam circulates on the outside of the tubes while the evaporating charge boils up inside the tubes and down wider openings provided for the purpose. An open steam coil is also provided in the bottom of the still for blowing off final traces of solvent at the end of the operation.

The oil-finishing still is very similar in construction to the solvent-recovery still, and where such installation is made its purpose is to complete the operation of removing the solvent left unfinished in the preceding solvent-recovery unit. Where the latter is used for the preliminary separation of oil and solvent, these stills, particularly the oil-finishing still, should be equipped for operating under vacuum.

The solvent and water separator is usually a tank with a separate compartment and overflow for drawing off the water which has settled from the distilled mixture of solvent and water. Such separation must be effected in a quiescent condition, because even a slight agitation of the layers precludes quantitative recovery. It may be necessary to install a preliminary water-solvent separator to take up the pumping pulsations in order to obviate the churning effect when the vapors and condensed liquor are discharged direct from the vacuum pump during any part of the boiling-off operation.

The other equipment of such a plant consists of storage tanks for saturated solvent and oil, for solvent alone, and for oil; also a solvent heater, boilers, and conveyors. All storage tanks should be provided with indicators to record the level of the contents, or when buried under ground with indicators above ground to register the contents.

Heretofore there has not been much inducement for manufacturers to prepare a high-grade No. 3 oil. Consequently, the generally recognized darkening effect of heat due to steaming off has not been considered. If, however, effort is made to produce a high-grade oil with a minimum of color, care must be exercised not to expose the oil in the solvent-recovery still or oil-finishing still to this high temperature.

By such solvent extraction, as generally practiced, a pomace is produced running from 10 to 12 per cent of moisture and containing anywhere from 0.6 to 2 per cent of oil. This low content of oil can be attained only by very efficient extraction, determined by the kind of solvent used, the number and efficiency of the washings, and the type of equipment.

SOLVENT-EXTRACTED NO. 1 OIL.

In the course of the work of this laboratory on the technology of castor-oil manufacture, evidence has been obtained which indicates that a very promising oil of apparently No. 1 grade can be made entirely by the solvent process. Without decorticating the beans, they were slightly crushed in order to break the seed coats and extracted with benzol by percolation. The solvent saturated with oil was of a characteristic light-green color, similar to oil always produced by such extraction. The benzol was evaporated at atmospheric pressure until toward the end of the distillation, whereupon a vacuum was applied. The residual oil with greenish yellow color was then heated to 95° to 110° C. at atmospheric pressure and treated with 5 per cent of fuller's earth with constant agitation for about 10 minutes, whereupon 2 per cent of decolorizing carbon was added and the stirring continued, while the temperature was allowed to fall slowly to about 90° C. At the expiration of about 15 minutes the oil was filtered and came through with a light straw color. Inasmuch as every castor-oil expelling plant must have an auxiliary extraction equipment, it follows that if a satisfactory grade of oil, colorless and low in acidity, can be made by such means it will materially reduce the cost of production.

TREATMENT OF POMACE.

The dry, more or less dusty pomace from the extraction house is conveyed to the pomace warehouse. If it is lumpy, it is passed through a grinding machine, which may be of the rotating-disk type, spiked-roller (crusher), or swinging-hammer type, where it is pulverized. If the moisture runs above 12 per cent, decomposition may set in, resulting in an appreciable loss of ammonia, the only constituent of real commercial value.

FINISHING THE OIL.

REFINING.

Various methods are recorded in the literature for refining castor oil. Among these may be mentioned acid refining by sulphuric acid, settling of the foots, and subsequent washing out of the acid. Other methods involve refining in alcoholic solution. As far as can be ascertained, these methods are not generally applied. British practice has long consisted in passing live steam into the oil, which serves to coagulate and precipitate the albuminlike constituents, which are filtered off. That this product is nitrogenous has been demonstrated. The oils listed in Table VII were analyzed for nitrogen.

TABLE VII.—*Analyses of castor-bean oils of diverse origin, showing nitrogen content.*

Source.	Nitrogen (N × 6.25) content (per cent).	
	Raw.	Treated.
American	0.0159 / .0169 } 0.0164	0.0163 / .0163 } 0.0163
Newchwang	.0399 / .0469 } .0434	.0175 / .0188 } .0182
"Chinese"	.0338 / .0363 } .0350	.0163 / .0183 } .0173

Castor oil can not be freed from acid as readily as some of the other staple oils, such as cottonseed and peanut oils. Instead of the soaps breaking readily and settling out, the castor-oil soaps only partly do so but have a tendency also to dissolve in the oil, rendering the latter very viscous and thick. Consequently, the oil has to be very thoroughly washed to remove such.

No. 1 castor oil is usually so low in acid that it serves all industrial purposes without refining. However, there are times when the oil may run high in acid, on account of the character of the beans received at a given plant, or the acidity may run up, owing to factory troubles. All No. 3 oil is high in acidity, with a range of possibly 5 to 7 per cent figured as oleic acid. All such oil can be refined as follows:

The oil should first be heated to about 85° C. and treated with caustic-soda solution approximately 16° Baumé. With the temperature maintained at this point the oil and alkali are agitated gently to insure intimate contact, whereupon agitation is withdrawn and the aqueous soapy layer allowed to separate as much as it will. It is then drawn off. The oil is then heated to around 95° C. and sprayed with successive portions of boiling-hot water (preferably brine) with thorough agitation. Agitation that is too violent produces a troublesome emulsion, which can only be broken with difficulty and by long heating or sweating out. If, however, the hot oil is given comparatively mild agitation over the layer of hot water which has settled to the bottom, with just enough motion to insure good movement between the two phases without emulsifying them, it is found that all soap can readily be removed. If an emulsion should form, it can be broken by heating the oil and sprinkling in salt, with such agitation as has been described. Upon the settling out of the brine, it is drawn off and the oil dried and filtered.

No bleaching has been effective for the regular commercial types of No. 3 castor oil. This is attributed to the fact that the present practice has fixed the color by overheating and at the same time has introduced iron salts, due to the action of the comparatively high acidity upon the container walls. It has been found that castor oil, though highly colored, can be refined by alkali treatment and bleaching only when the oil has been treated with the same care that applies to any other oils intended for bleaching. A greenish yellow extracted oil has been successfully bleached, as well as a green oil produced by expelling castor beans in their original hulls or pods, but no success has been attained in endeavoring to bleach any of the commercial types of No. 3 oil, whether green or brownish yellow.

BLEACHING.

Castor oil coming directly from the presses or expellers is of a brownish gray color, due to suspended meal and droplets of water. Inasmuch as the oil produced by cold pressing or expelling runs low enough in acid to be satisfactory for the general purposes for which No. 1 castor oil is intended, it is not refined in practice, but merely cleaned up and bleached. Before satisfactory filtering and bleaching can be accomplished the oil must be dried. This is effected by heating the oil either at atmospheric pressure or in vacuum. Such equipment consists of an oil tank with closed coils, agitator, sight glasses, and a small 4-inch entrainment, which seems small but is ample to carry off the comparatively small amounts of water. Closed steam coils supply the necessary heat. Effective agitation is necessary to bring the moisture-bearing portions to the surface for ready drying. The oil may be circulated by pumping from the bottom and in at the top in the form of a spray, which, of course, operates to minimize the period of heating. After such drying is effected the oil is pumped to a mixing kettle and treated with fuller's earth and carbon. The practice differs in different plants and according to the grade of the oil. A common method is to heat the oil to approximately 200° F., preferably in vacuum, and then add from 2 to 4 per cent of a good grade of dry fuller's earth. This mass is agitated for half an hour and then a good grade of bleaching carbon is introduced, varying from 0.2 to possibly 1.5 per cent, according to the grade of oil and quality of product desired. The oil is filtered and is then ready for market.

A centrifugal machine (fig. 14) has been devised departing in some respects from the ordinary basket or cone centrifuge, which has been found quite satisfactory on a laboratory scale for removing from the oil large quantities of the sticky albuminlike product which on long standing settles from oils and which so seriously impairs the capacity of the filter press. This machine by removing such meal and other substances as are difficult to filter out leaves an oil with only a thin cloud, but which can be treated with fuller's earth, bleached, and filtered with much greater ease than with the bulky precipitate retained. The walls of this centrifuge are solid and not perforated, as in the laundry type. A series of baffles (fig. 15) attached alternately to the wall and to the central shaft forces the oil to take a deviating course through the centrifuge, resulting in the collection of the precipitate in the lower portions of the centrifuge somewhat similar to the throwing out of moisture from steam

FIG. 14.—A centrifugal separator.

when the latter passes through the trap or catchall in a vacuum entrainment. (Fig. 11.)

The hot oil with its charge of fuller's earth and carbon is then pumped through an ordinary plate-and-frame filter press, the cloths of which should be covered with paper in order to obviate the trouble and difficulty of washing out the adhering earth from the pores after each filtration. All that is necessary in this case is to wash the cloths with alkali after several runs in order to remove the oil without the attendant rubbing necessary to clean up the pores. The filter papers with their charges of earth and carbon can be lifted off the cloths and thrown into the solvent extractors, thus greatly minimizing the labor of cleaning up the press. As is usual in such operations, the oil must be pumped through the filter press and back into the original mixing tank for half an hour or so in order to coat the leaves with a layer of earth, owing to the fact that the first runnings of oil are cloudy and clear up only when the plates are well covered. The stream of clear oil should then be pumped through a

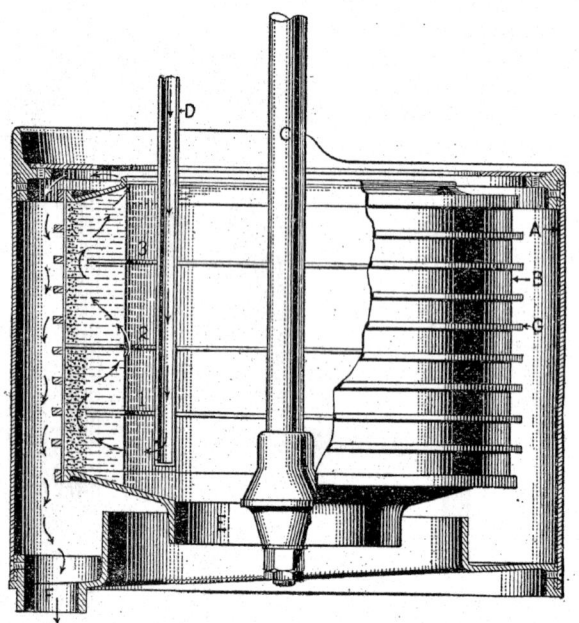

FIG. 15.—Sketch showing details of construction of the centrifugal separator shown in figure 14.

second press in order to catch any fine specks which may escape the first filtration.

Such treatment should produce a very light oil, no darker than a fine straw color. However, low-quality beans and faulty factory practice may result in greater color and unreasonable acidity. Factors contributing to high acidity are moisture and heat, particularly the former, while those affecting color are largely heat, particularly when the oil is heated in contact with seed coats or hulls. A clear bleached oil heated for a little over half an hour at 195° F. materially darkens, but samples of oil heated for one hour at 400° F. suffered practically no change in acidity. This refers to both No. 1 oil and refined No. 3 oil. The acidity of the oil does not materially increase during the several days which elapse between manufacture and finishing. Under common practice the oil from the first day's run is collected at the end of the day for weighing. It is then dried over part of the following day and the third day is finished by bleaching and filtering. During this time the increase in the acidity is almost negligible, increasing only about 0.1 of 1 per cent. An oil can hold 0.15 per cent moisture without clouding, but when the moisture runs up to as much as 1 per cent the moisture settles out and a permanent cloud is formed.

PROPERTIES OF THE OIL

The quality of castor oil is determined in the trade by its color, clearness, and acidity. The best grade, designated as No. 1, is a cold-pressed oil low in acidity, brilliantly clear, and approaching colorlessness. Some cold-pressed oils are a very light straw color, almost white; all hot-pressed oils are brownish yellow, while the so-called No. 3 grade varies from dark brown to green. There is no generally recognized No. 2 grade, since this was designated by only a few dealers to represent some stock inferior to No. 1, but better than No. 3. The specifications accompanying trade in this commodity usually consist of clearness, color, acidity, specific gravity, iodin value, saponification value, odor, and taste.

TABLE VIII.—*Analyses of commercial samples of castor oil covering the period from November, 1917, to December, 1918, and of expeller oils made at the Government plant at Gainesville, Fla., in the spring of 1919.*

Properties compared.	Hydraulic-pressed oil.		Expeller oil (series of more than 100 samples).	
	Range.	Average.	Range.	Average.
1	2	3	4	5
Specific gravity (15.6° C.)	0.961 to 0.965	0.9626	0.9610 to 0.9634	0.9620
Acidity .... per cent..	a 0.45 to 1.6	1.22	0.60 to 1.83	.98
Iodin number	83.0 to 87.5	85.6	b 82.7	
Saponification number	176.3 to 188	182.5	b 181.4	
Unsaponifiable matter .... per cent..	0.21 to 0.78	.33	b 1.0	
Flash point .... ° F..	513 to 570	535		
Viscosity	a 93 to 112	97.1		

Properties compared.	Expeller oil (series of 7 samples).		Extracted oil.	
	Range.	Average.	Range.	Average.
1	6	7	8	9
Specific gravity (15.6° C.)	0.961 to 0.962	0.9613	0.9607 to 0.9624	0.9614
Acidity .... per cent..	0.85 to 1.46	1.17	a 3.09 to 6.66	4.80
Iodin number	80.8 to 86.1	84.2		
Saponification number	178 to 182	179.6		
Unsaponifiable matter .... per cent..	0.20 to 0.30	.26		
Flash point .... ° F..	524 to 527	525		
Viscosity				

a One higher value in each case was discarded as unrepresentative.   *b* One sample.

32   BULLETIN 867, U. S. DEPARTMENT OF AGRICULTURE.

An average of 54 analyses of commercial samples of No. 1 castor oil manufactured by the leading crushers, covering the period from November, 1917, to December, 1918, is given in Table VIII, columns 2 and 3. These hydraulic-pressed oils are from cage presses and represent the normal product regularly made. In columns 4 to 7 similar data are given for expeller oils made at the Government plant at Gainesville, Fla., including more than 100 samples manufactured during the spring of 1919, and in columns 8 and 9 are given the specific gravity and acidity of samples of extracted oil.

That the refining of castor oil does not materially alter its chemical characteristics other than merely to remove the acid is shown in Table IX. Sample A represents a refined No. 1 oil the acidity of which had risen to 10 per cent, while sample B is from a batch of No. 3 oil chemically refined.

TABLE IX.—*Comparison of properties of representative samples of castor oils differing in acidity.*

Properties compared.	Sample A.	Sample B.
Specific gravity	At 15.25 Bé. 0.9643	At 15.70 Bé. 0.9613.
Flash point °F.	555	555.
Fire do	630	615.
Pour test do	0	0.
Viscosity at 212° F seconds	100 to 101	96 to 97.
Ash	(1)	(2).
Total saponification	180.85	182.48.
Odor	Castor	Castor.
Water	Trace	Trace.
Appearance	Light yellow, pale	Light green.
Iodin value	87	91.
Solubility	O. K.	O. K.

[1] Trace, considerable alkalinity, but no sulphates. [2] Trace, slight trace of alkalinity, but no sulphates.

The following is an analysis of a single sample of No. 1 expeller oil containing properties not determined in the foregoing analyses:

Color	Straw, very slight cloud.
Odor	Mild.
Specific gravity at 15.6° C. (60° F.)	0.963 = 15.4° Bé.
Solubility in 90 per cent alcohol	Completely soluble.
Acid number	1.9 = 1 per cent free acid (as oleic).
Iodin number	82.7.
Saponification number	181.4.
Viscosity	98.6.
Unsaponifiable matter (per cent)	1.0.
Rosin	None.
Adulterants	None detected.

Samples of castor oil from various sources were heated to 400° F. for 1 hour to ascertain the effect of heating on the acidity. These results are given in Table X, and are reported as the percentage of acidity expressed as oleic.

TABLE X.—*Analyses of castor oils from various sources, showing the effect of heating upon acidity.*

Description of sample.	Acidity (per cent).	
	Control.	Heated.
Refined No. 1 oil	0.47	0.58
C. P. No. 1 oil	1.00	1.05
Refined No. 3 oil	.71	.79
Refined No. 1 oil	.92	.96

It is thus evident that heat alone has little effect on the acidity. This question arose from the possibility that if such acidity did develop, it might operate to pit the walls of the gas-engine combustion chamber during a run.

When the Bureau of Aircraft Production went into the market for castor oil for lubricating purposes it drew up the specifications listed below, giving the properties which a good grade of lubricating castor oil should possess. Since these properties are possessed only by a high-grade No. 1 oil, these specifications may be accepted as fairly representative of this entire grade, regardless of its intended use.

General.—(1) This specification covers the requirements of the Bureau of Aircraft Production in all purchases of castor oil for rotary-engine lubrication. The oil must be a high-grade vegetable castor oil suitable for this purpose. Both cold-pressed vegetable castor oil and hot-pressed vegetable castor oil which has been refined so that it will meet the requirements of this specification may be submitted for purchases. (2) The castor oil must be free from adulteration, other oils, suspended matter, grit, and water. (3) The castor oil must meet the following requirements:

Color.—(4) When observed in a 4-ounce sample bottle, the castor oil must be color less or nearly so, transparent, and without fluorescence.

Specific gravity.—(5) The castor oil must have a specific gravity of 0.959 to 0.968 at 60° F. (Baumé gravity must be from 16.05 to 14.70 at 60° F.)

Viscosity.—(6) The castor oil when tested in a Saybolt universal viscosimeter must have a viscosity of not less than 450 seconds at 130° F. and 95 seconds at 212° F.

Flash point.—(7) The flash point must not be less than 450° F. in a Cleveland open-cup flash tester.

Pour test.—(8) The castor oil, in a 4-ounce sample bottle one-quarter full, must not congeal on being subjected to a temperature of plus 5° F. for one hour. (See specification No. 3525, "Pour test.")

Evaporation test.—(9) The castor oil must not show a greater loss than five-tenths of 1 per cent when heated in an oven at 230° F. for 1¾ hours. This test shall be made on a 5-gram sample in a glass crystallizing dish approximately 2½ inches in diameter and 1½ inches high, inside dimensions.

Ash.—(10) The castor oil shall not show more than 0.015 per cent of ash and shall show no impurity of any sort not related to the original product.

Solubility.—(11) The castor oil must be completely soluble in 4 volumes of 90 per cent alcohol (specific gravity 0.834 at 60° F.). This test shall be made on a 2 c. c. sample.

Acid number.—(12) It must not require more than 3 milligrams of potassium hydroxid (KOH) or 2.14 milligrams of sodium hydroxid (NaOH) to neutralize 1 gram of oil. This is equivalent to 1.5 per cent of oleic acid.

Unsaponifiable matter.—(13) The unsaponifiable matter must not exceed 1 per cent. Samples used for this test shall weigh 5 to 10 grams.

Iodin number (Hanus or Wijs methods).—(14) The iodin number must be between 80 and 90. Samples used for this test shall weigh 0.2 to 0.25 gram and shall be treated for 1 hour.

Rosin (Lieberman-Storch test).—(15) The castor oil must not give a reaction for either rosin or rosin oil.

Cottonseed oil (Halphen test).—(16) The castor oil must not give a reaction for cottonseed oil.

Inasmuch as the chemical analysis does not give the final word regarding the adaptability of an oil for lubricating purposes, engine tests have been made on the lubricating value of No. 1 hydraulic-pressed oil, No. 1 expeller oil, and No. 3 refined oil. The results of such tests show these oils to be of equal value for lubricating purposes. Since the chemical and physical constants expressed above are practically identical, it follows that color is the only evident means of differentiating between the various oils. A demulsibility test applied to hydraulic-pressed oil compared with expeller oil reacted slightly in favor of the expeller oil. The difference, however, was so slight that the two oils may be considered in this respect practically identical.

The following statements quoted from leading dealers in castor oil (not manufacturers) show how the trade considers American-produced oil as compared with various imported stocks:

The American-pressed castor oil will remain free from rancidity for a longer period than the imported oils and as a general average is vastly superior to any imported oil that we have received.

The oil that comes from China and the Far East seems to be of a decided yellow color and in the writer's judgment, would indicate that it is hot pressed, i. e., that the oil was pressed from a warm or hot meal.

In our opinion the oil made in the United States is equal, if not superior, to the imported.

We will say that it has happened that the oil we purchased which was made in this country turned out to be better than that we have used which was made abroad.

Domestic-manufactured castor oil will keep longer and be freer from acidity than the oil which is imported. Generally speaking, the imported castor oil, especially from the Orient, contains from 1 to 8 per cent acidity and by keeping the oil the acidity is likely to be increased, especially where the oil tests from 3 or 4 to 8 per cent. We look upon the domestic-manufactured white oil as being best not only for medicinal but for manufacturing purposes.

As a matter of fact, now that the War Trade Board has ruled (January, 1919) that castor beans and castor oil can come freely into this country, we doubt if any of this oil (oriental) will come here. It is, as you may know, an inferior oil, and can only be used in comparison with a domestic production of No. 3 castor oil.

USES OF CASTOR OIL.

Castor oil has properties which serve to differentiate it very markedly from all other vegetable oils. This fact is no doubt attributable to the predominating influence of its characteristic acid radical of ricinoleic acid. The striking properties of this acid are due to the fact that it is a hydroxy acid, a condition which is only approximated in nature in the case of grape-seed oil. This hydroxylated condition is probably the property which serves to render it so valuable in the industries, for it is apparent that when other oils are treated so as to increase their acetyl number and their viscosity they more nearly take on the properties of castor oil. Thus, when an oil is oxidized (blown) it becomes less soluble in gasoline, and its viscosity and acetyl value increase. In fact, the hydroxylation of acids normally soluble in gasoline renders them insoluble in gasoline. Sulphonated oils are insoluble in gasoline. Since hydrolyzing a sulphonated oil is said to yield hydroxylated oils, efforts have accordingly been made in this and other laboratories to produce such an oil, using peanut and cottonseed oils, but without success thus far.

For ordinary lubrication, the viscosity of castor oil is its great asset. Before the application of mineral oils for such purposes castor oil was very largely used as a cylinder oil, but the production of high-grade mineral cylinder oil has greatly displaced it except in the Tropics, where it is still used for lubricating heavy machinery. However, in gas engines, which are lubricated by spraying the lubricant into the cylinder along with the gasoline (notably the rotary air-cooled types), it has been found that castor oil is absolutely necessary. The property of the solubility of mineral oils in gasoline is stated to be the reason that they can not be so used as a lubricant, owing to the lowering of their viscosity and "body" by solution in such a medium.[1] The insolubility of castor oil in such products is given as the cause for its specific advantages in such cases. Some authorities assert that castor oil is preferable to mineral oil because gasoline does not wash it out of the crank case so readily, which is, of course, a corollary of the above. However, mixtures of cylinder mineral oil and castor oil treated so as to maintain their homogeneity are stated after direct trial to be the more satisfactory, although such mixtures are perfectly soluble in gasoline. Castor oil, of course, can be used as a lubricant in other types of motors, but as the supply has been

---

[1] Although the literature almost universally states that castor oil is insoluble in gasoline, attention is called to the fact that this qualitative statement should be restricted to refer only to conditions obtaining at ordinary temperatures. A gentle heating effects ready solution; in fact, extraction of castor-press cake pomace with gasoline is the industrial method for obtaining the lower grade oil, leaving a pomace with about 2 per cent of oil. In view of the high temperature conditions obtaining in engines lubricated with castor oil, it is apparent that this oil readily dissolves in gasoline; consequently, its specific advantage for lubrication under such conditions would not appear to reside in its insolubility in gasoline but rather to the fact that solution in gasoline leaves it with a viscosity higher than that obtaining with other oils similarly treated.

somewhat limited and at the same time the price materially higher than that of high-grade mineral oil it was found that the latter can be used in ordinary types, while only the former can be used in the rotary types. Abroad, however, castor oil was used almost exclusively in all kinds of aviation motors. The reason given was that castor oil keeps its viscosity better, sticks better, and protects the cylinder walls, valve seats, and other parts. A mixture of mineral and castor oils, containing a preponderating percentage of the latter, has been universally used in stationary motors. Some trouble with foreign castor oil has been due to its tendency to gum, which has been minimized by mixing it with various proportions of heavy mineral oil. Great difficulty has been experienced in forming a homogeneous and suitable mixture of castor and mineral oils, owing to the fact that both are apparently homogeneous at the time of making, but separation occurs upon long standing. A patent has been taken out by Archbutt and Deeley for heating castor oil in an autoclave at 260° to 300° C., under pressure of 4 to 6 atmospheres for about 10 hours, whereby it becomes miscible with mineral oil in any proportion. Some state that castor oil unduly precipitates carbon on the walls of gas-engine cylinders, owing to incomplete combustion, while others claim that because castor oil burns without a smoky flame and gasoline burns with a conspicuous cloud, any carbon deposited on the cylinder walls is derived from the gasoline rather than from the oil.

Some lubricating engineers claim that the high steam pressures occurring in steam cylinders afford ideal conditions for saponifying castor oil, stating that this increases the acidity of the oil, with consequent pitting of the walls. Others claim that the walls remain perfectly bright. Another quality of castor oil as a lubricant, which seems to be quite generally accepted, is its ability to stick to the exposed surfaces, with consequent protection. Castor oil also keeps its viscosity better under changes of temperature than any other vegetable oil and many mineral oils.

Artificial leather is made by dissolving cellulose nitrates in volatile solvents, incorporating castor oil in the mixture and distributing the same over treated cloth. Upon volatilization of the solvent, the solid constituents remain fixed on the goods. The rôle of the oil is to impart softness and elasticity to the otherwise hard and stiff product and to enable this to be more readily coated on the cloth or other backing material. There are very few oils which can be added to nitrocellulose solutions without either causing the separation of the nitrocellulose from the solution or spoiling the luster and cohesion of the film. It is evident that a nondrying oil must be used and also one that is perfectly miscible in the solvents used. Castor oil fulfills these conditions very satisfactorily and is of additional value on account of its resistance to climatic conditions and temperature changes as well as to its viscosity. A leather substitute, recently patented, is formed of a carrying vehicle, such as paper or a woven fabric, and a facing of supple pyroxylin built up of successive layers united into an integral structure of sufficient thickness to enable it to be removed from the carrier. The coating may be formed of nitrocellulose 10, castor oil 20, amyl acetate 15, methyl alcohol 20, amyl alcohol 5, benzol 30, and pigment 3 parts. Such leatherlike products come in rolls of 30 to 60 yards in length and of varying widths, and find extensive use in upholstery, carriage tops, automobile fittings, suitcases, trunks, shoes, book bindings, and various lines of novelty goods. It has been generally assumed that only the No. 1 grade of castor oil is satisfactory for this purpose, but progressive manufacturers have learned that a properly refined No. 3 oil, although it runs high in color, can readily be used, inasmuch as most artificial leather products are of dark color. As is evident from the analyses previously recorded, the characteristics of the oil after refining are in no wise deleteriously affected.

In the leather trade castor oil finds rather extensive use both as a lubricant and as a soluble oil. Specifically, it is applied to belting directly as a sulphonated product and is also incorporated in a composite grease which may contain in addition to the oil such products as tallow, wax, paraffin, and vaseline. Belts treated with this mixture are made flexible and are prevented from cracking, all of which operates to increase the friction on the pulley.

It is stated that castor oil applied to leather in snowy weather keeps the leather soft and makes it waterproof; also that leather so treated is avoided by rats. It does not prevent a polish being produced on boots, and if applied once a week to leather shoes will cause them to last twice as long. Such treatment is particularly recommended if the leather has been wet. In such cases the oil should be rubbed in before the goods have dried. The softening of leather belts, harness, and other such leather goods is a further use to which castor oil is put.

Sulphonated castor oil is made by treating the oil with sulphuric acid under carefully controlled conditions of temperature and proportion of ingredients. The resulting product may be soluble or readily emulsifiable in water. It possesses the property of emulsifying other oils and greases and carrying them into the leather, which thereby becomes lubricated internally. Mineral oil may thus be carried into leather and impart to it a certain interior humidity. Sulphonated castor oil also facilitates the penetration of tannin into leather. It forms an ingredient of various creams, both black and colored, for rubbing patent leather.

Sulphonated castor oil is the basis of manufacture of emulsifying or soluble cutting oils used in connection with water. It may also be used with mineral oils as a cutting oil when no water is used. It appears to have greater cooling qualities than most vegetable oils and does not gum or become rancid.

Sulphonated oils, particularly castor oil, have been used in producing the dye called Turkey red. Cotton cloth is treated with alum and immersed in a bath containing a solution of sulphonated oil (soluble in water). This is thought to form an aluminium oleate, which acts as a mordant to form a lake. Treating the mordant cloth with alizarine results in a bright red lake called Turkey red. The relative quantity of this color that is now used is considerably smaller than formerly, owing to the use of other colors of a similar shade.

For mantle dips this oil, as well as other vegetable oils, is used as a softener to render the coating of the mantle flexible. After the mantle has been dipped the coating on it has about equal proportions of cotton and the material from the oil.

In the manufacture of linoleum, castor oil has been found to be of advantage in imparting flexibility and toughness to the goods, somewhat the same as in imitation leather. Both No. 1 and No. 3 oils have been used, but since the finished goods are usually somewhat colored, no reason exists why an acid-free (refined) No. 3 oil would not be perfectly satisfactory.

Vegetable oils, notably castor oil, may be treated with sulphur and vulcanized, similar to rubber. This may be effected either by treating the oil with sulphur chlorid or by fusing it with sulphur direct. In the first case the product is known as "white substitute," due to the comparatively light color of the product, while in the second process the product is known as "black substitute." Sulphur chlorid may be added to the oil direct in proper equipment to control the temperature, or it may be added to a solution of the oil in some solvent. In either case the mixture, more or less hot, may be poured into molds or cooled, then ground and dried. The same treatment is pursued in the case of both the black and the white products. On heating, both products mix well with rubber; hence the name "substitute." The rôle of this product does not necessarily have to follow that of an adulterant in the sense that it is a mere cheapener. Certain rubber goods are not satisfactory unless mixed with other products. For example, the specific gravity of vulcanized oils is lighter than that of rubber and their incorporation offsets the increased weight due to mineral filler. They also impart softness to the product, desirable in certain fabrication.

It has been stated that so little sulphur chlorid is necessary for vulcanizing castor oil to make it set that it is difficult to work with. Some authorities state that castor oil must be used for floating substitute.

Among other uses in the rubber industry, castor oil finds application in the manufacture of gas tubing, insulating tape, and packing sheets.

Cellulose nitrate "dope" is greatly improved by the addition of 5 to 7 per cent of castor oil or treated tung oil. Greater elasticity of film and slow evaporation result.

Castor oil lends elasticity to varnish and has been stated to be an ingredient in certain artificial skin preparations, the formula for one of which is shellac, 1 part; alcohol, 3 parts; castor oil, one-fifth part. It is also used in retouching varnishes and in photographic-negative varnishes. In general, its use in varnish is to lessen brittleness and minimize the attendant property of chipping and peeling.

Castor soap is transparent, white, and quite hard. It dissolves in cold water without rendering the latter turbid. It lathers quickly and is very soluble.

In the manufacture of tire cement castor oil forms an ingredient of good thick shellac varnish. It prevents the bicycle rim from becoming hard and brittle.

Many salts of the aniline series are soluble in castor oil and advantage has been taken of the fact to prepare typewriter inks of great copying power, which permit large numbers of copies to be taken from the same impression. Hopkins [1] states that such inks are very little affected by extremes of dryness, moisture, heat, or cold. He also states that the oil-soluble colors are not affected by the moisture of the hand. Castor oil also prevents the ink from drying on the pad and at the same time "bites" the oil-soluble aniline color into the paper and prevents it from rubbing. Objection, however, to the use of the oil is that impressions from such inks are often surrounded by greasy marks caused by the fats spreading in the pores of the paper, and that the present practice is to make most of the stamping inks without grease by preparing mixtures of coal-tar dyes in glycerin.

In the manufacture of fly paper castor oil is a necessary ingredient. Various combinations of castor oil, resin, and other products are spread upon heavy paper with a common glue sizing. It is stated that sugar is sometimes used to make the product more attractive.

By heating nitrated oils to 130° C. or by oxidizing them with lead peroxid, rubberlike substances are obtained. Nitrated castor oil, made by nitrating with a mixture of 2 parts sulphuric acid and 1 part nitric acid, finds use in industry through its property of making homogeneous compounds with nitrocellulose. Such a mixture yields a product resembling ebonite. Solutions of nitrated oils in acetone are used as varnishes, as a basis for paint, and for enameling leather.

Castor oil finds a further use in the textile industry as a so-called "wool oil" (sulphonated castor oil), and very commonly is referred to as "castor-soap oil," both of which are used for degreasing special woolen products.

---

[1] Hopkins, A. A., ed. Scientific American Cyclopedia of Formulas . . . 1077 p., illus. New York, 1911.

The potassium soap of this oil is used as a solvent for water-insoluble substances, as the ethereal oils, cresols, and synthetic perfumes.

The extensive use of castor oil in medicine is due to its purgative property. Just to what this is due is a mooted question. Some think that the presence of small amounts of ricine or some other impurity imparts to the oil this property, which is lacking in the pure glycerid or oil. Others claim that this property is characteristic of ricinoleic acid (the acid radical of the oil) and quote in support of their contention the fact that pure ricinoleic acid itself is purgative. On the other hand, the statement that castor oil extracted from the seeds by alcohol is more effective than that made by expression lends color to the belief that the solvent plays a selective rôle in extracting more of the substance which possesses the purgative property.

Much effort has been expended in attempting to remove from castor oil that property which makes it so repugnant to the taste and smell. Simple deodorization in a vacuum deodorizer is not altogether satisfactory. Everyone is conversant with the corner druggist's effort to mask it in soda water, peppermint, and other "sandwiches." Coloring it and adding a tincture of some of the common spices is about as satisfactory as any method. The following rather unique concoction is quoted (J. King, King's American Dispensary) for the reason that if such an unpleasant product as castor oil can be made to simulate a custard, even remotely, the fact should be made known to all:

I find it a very pleasant mode of administration to boil the dose of oil with about a gill of good sweet milk for a few minutes, sweeten with loaf sugar and flavor with essence of cinnamon or other favorite aromatic; it somewhat resembles custard in its taste and appearance and is readily taken by even the most delicate stomach.

CONCLUSIONS.

It is thus seen that although castor oil is one of the minor oils, its industrial use is increasing in a marked degree. While it is more widely known for its medicinal properties, its use is being constantly extended in a variety of industries.

The general method of manufacturing the oil in this country has been by crushing the beans in cage presses, but it has been found that the expeller produces an oil of satisfactory quality for all industrial uses and is perfectly satisfactory for aeroplane lubrication. Evidence has been obtained that a good grade of No. 1 oil can be obtained by extraction with volatile solvent. Highly acid dark oil can be refined by alkali but not highly bleached, while low acid oils can be refined and bleached to almost water white. Attention is called to the varied uses made of the oil and the possibility of finding markets for the more sluggish No. 3 oil.

The castor-oil industry
Shrader, J. H (James Houston), 1885-
U.S. Department of Agriculture, National Agricultural Library

CPSIA information can be obtained
at www.ICGtesting.com
Printed in the USA
LVOW01s1137040517
533239LV00017B/511/P